WORLD WITHIN A WORLD

Everglades

WORLD WITHIN A WORLD

Everglades

Written and illustrated by
TED LEWIN

With an Introduction by Don R. Eckelberry

DODD, MEAD & COMPANY · New York

To Wild Places

Illustration on page 1, wood stork; title page, turkey vulture on moon vine

Library of Congress Cataloging in Publication Data

Lewin, Ted.
    World within a world—Everglades.

    Includes index.
    1. Natural history—Florida—Everglades. I. Title.
QH105.F6L48   574.9759'39   76-7361
ISBN 0-396-07329-8

Contents

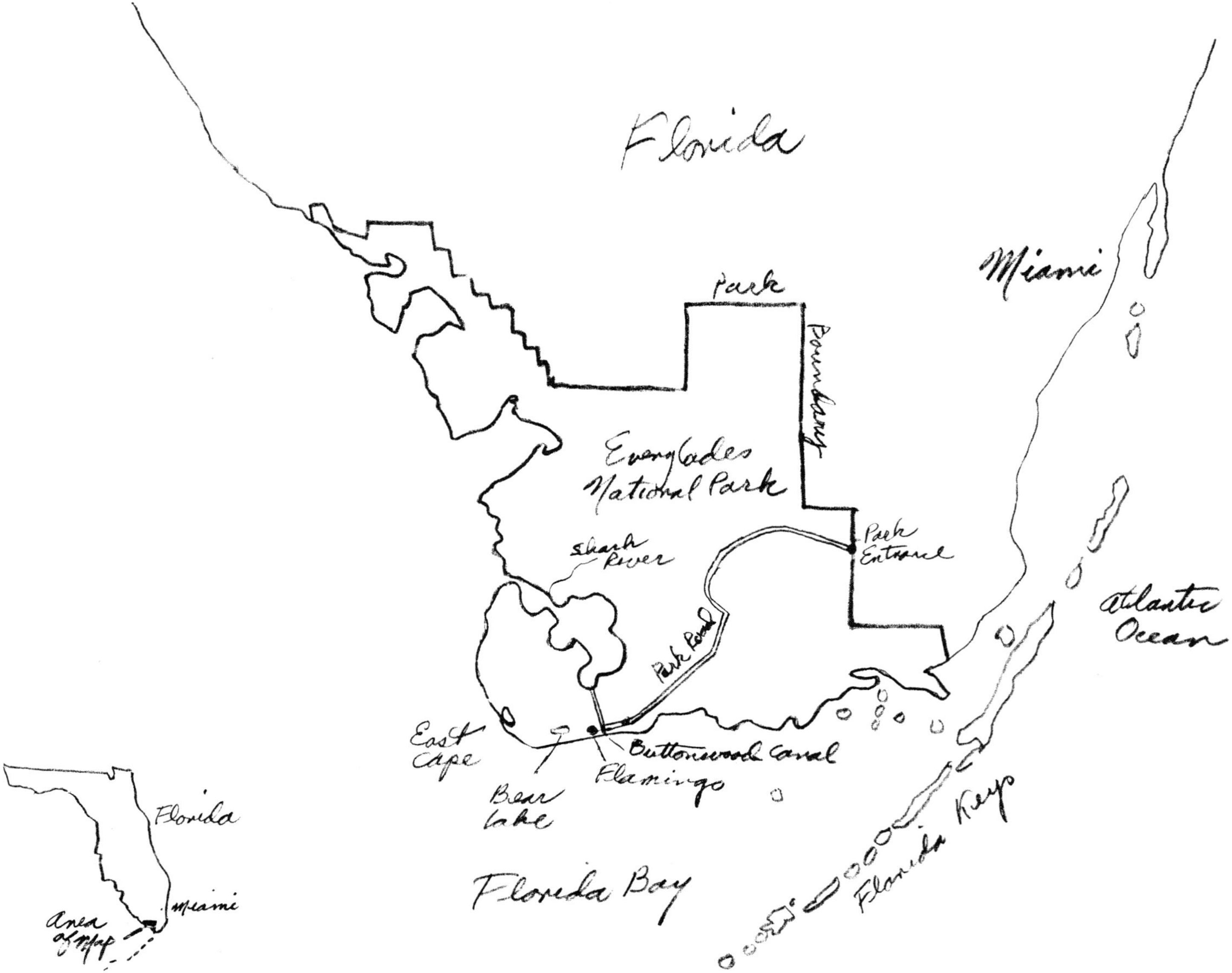

Florida

Miami

Park

Boundary

Everglades
National Park

Shark
River

Park
Entrance

Park Road

East
Cape

Buttonwood Canal

Flamingo

Bear
Lake

Atlantic
Ocean

Florida Keys

Florida Bay

Florida

Miami

area
of map

Introduction

The strong delicacy, the refined simplicity and the style of Ted Lewin's drawings need no comment from me, as these qualities may be immediately appreciated. It is difficult to say whether his terse and cleanly written stories are meant to bring the pictures together or the other way round, so much do both express the same vision. The main thing is that there is not just truth but *life* here, or to put it better—*aliveness.* This book is not the result of a casual tour of one of our great biological areas, but rather it is the creative condensation of five years of canoeing, tramping, and camping in the Everglades.

DON R. ECKELBERRY

Limpkin - Everglades 1975

1.
The Limpkin

The world of the Everglades is an event, or more precisely, a series of events. Subtle, violent, quiet, raucous, ceaselessly unfolding and seemingly diverse, the events are interrelated parts of the whole, each leading irrevocably to the next.

A novice observer, I went to the Everglades because I became aware that this unique wilderness area was in danger. I found a fragile world that has, for centuries, been assaulted by natural forces and, more recently, by man. But part of it is still there, and I have returned often since that first visit.

Come with me now as I journey through the Glades, from the saw grass marsh to Florida Bay. We'll share some of the things that draw me back so often to this enchanting place.

I start my search a mile or two from the entrance to Everglades National Park, just off the road, walking a path through the saw grass. The underfooting is wet and soggy. Wet is the key.

There are snail shells beneath my feet . . . apple snails, dependent on the moisture underfoot for survival . . . apple snails for the limpkin, shy elusive marsh bird so dependent on these snails for food.

A few more steps . . . a crying wail, a rush of wings, the series of events quickening . . . the limpkin, lifting out of the marsh where he has been feeding. A weak flyer, he settles in a tree a short distance away.

Twenty feet farther, screaming red-shouldered hawks swoop down, disturbed in their daily hunting by my presence. In my search for events, I've become an event for them.

The limpkin drops out of the tree and disappears. The hawks issue final angry screams and are gone. The pond is quiet. I walk on.

The pool is pristine. Just beneath the surface, a largemouth bass guards her mass of eggs from a hungry turtle . . . or perhaps herself from an alligator. Nearby I see the carapace of a large mud turtle. It is empty, picked clean. In the saw grass, an eagle feather . . . and my mind re-creates a long-finished event: the magnificent bird preening after his meal.

I leave, moving back along the same path. The bass still protects her spawn; no limpkin now, just the remains of his dinner. Wetness underfoot, wetness without which none of this could be. I wonder at the countless events that have transpired over the centuries . . . and more pointedly, how much longer these events can continue.

Laughing Gulls
Everglades
_____ 1975

2.
Snake Birds and Water Turkeys

The anhinga, a truly wonderful bird, needs two separate environments to survive. Snake like—and therefore called a snake bird—it has a long neck and scalelike feathers that hint at an ancient, reptile heritage. But, when swimming under water, its blue-black color and fanned tail make it look like a turkey gone mad—and so it is also called a water turkey.

I watch a young female fishing one day. She is rather inept, but fun to watch, anyway. She moves stealthily in and out of the spatterdock that covers the surface of the pond, looking for small bream or whatever she can spear with her needle-like bill. When she dives I can see her swimming beneath the surface. Occasionally she disturbs a huge mud turtle that turns and paddles slowly away.

Feeling the need to breathe, the anhinga pops to the surface, head and neck a feathered periscope, moving from one environment into the other. A breath, a quick arch of the

Anhinga - Everglades   [signature] 1975

neck, and she dives. The performance is repeated and repeated. I begin to fear she will catch nothing this day.

Suddenly, in the watery jungle below, two large bass flee from her. A moment later her head and neck break the surface. Impaled on her harpoon is an immense garfish.

Now the show for which I have been waiting—the quick flip of the head; the fish somersaulting in the air; the deft catch, headfirst for easy swallowing!

The anhinga flips her head and looks surprised when the gar, wiggling furiously, remains on her nose. A harder flip—still there. She climbs out of the water onto a low-hanging snag. Her big, flat, webbed feet wrap around the snag like wet washcloths, and the water beads and rolls off her blue-black feathers. The struggling gar is jeweled in the sunlight. The anhinga looks down her nose at her reluctant dinner. What now?

Slowly the spatterdock across the pond begins to gather and move toward the low branch where the anhinga sits with her dilemma on her nose. More and more spatterdock is gathered in, until the movement stops completely beneath the unwary bird with her unhappy garfish.

Hidden beneath the surface, a huge alligator wears the spatterdock like a deadly bridal veil. A neat trick! Or maybe just a curious accident. What is the gator after—fisherman, fish, or both?

The anhinga gathers herself for a final try. Feet tensed, neck cocked, she makes a great effort and up goes the fish. A flip—and the fish drops headfirst into the alligator's waiting jaws! The anhinga looks puzzled for a moment; then spreads her wings in the warm sunlight.

She will fish again. And the alligator waits.

3.
The Blue-winged Teal

Someone told me once that alligators never feed during the day. I should have known better than to believe it.

One morning at first light I watch a small armada of blue-winged teals. They have formed up at the other end of the pond and are proceeding (on whose orders, I don't know) to cruise toward me, their heads dipping in and out of the water. They are making quite a stir in the early morning quiet. They reach my side of the pond, turn, regroup, and head back into the reflection of the sun, which is cascading now into the pond.

A great blue heron stands in the center of the pond. He is absolutely motionless in the water, very seriously at work. There is a blur of white and slate-blue, a stiletto gleams, and his prize flaps, sending golden water droplets everywhere in the now warming sunlight. A calculated flip, and the heron's neck inflates to three times normal size, bulging into the shape of the catch. A gulp, and the neck returns to its lovely, slender S curve. The heron settles down to preen.

The teals are on their way back across the pond, sweeping their breakfast before them. The last duck in the line, a lovely little female, is having trouble keeping up. While I am watching her, it happens. I can't believe my eyes—they must be playing tricks on me! She is gone! Circles are spreading on the surface of the water. Did she really make them, or did a cormorant dive in the spot where she was swimming moments ago? There are cormorants about, but none reappears to breathe on the surface of the pond.

The teals come on relentlessly, now minus one. I continue to scan the surface carefully, using my binoculars . . . nothing.

Later that afternoon I visit the pond again. I still feel the unfinished business of the missing teal. I need an answer.

The scene is much the same, perhaps a little sleepier. The teals are still at work. I wonder at their appetites. I search the shore with my binoculars—the mudbanks, marsh grass, fallen snags. Something odd catches my eye—a pair of webbed feet are sticking out of the water at a peculiar angle. They are lifeless. Beside them, two eyes sit on the still surface of the pond—cold, emotionless eyes.

I have my answer. The alligator had slipped from his vantage point, his powerful tail propelling him forward silently. Settling down into the mud, he had waited like a killer U-boat—he was in his own kind of shipping lane. The unwary fleet passed overhead and, in an instant, one was subtracted. The gator slipped back down into the ooze, the duck—struggling for air, for life—held tightly in those terrible jaws. He moved away, finally,

and surfaced near shore under some overhanging branches.

The teals have no memory of the incident and several times they come within inches of the sated alligator. The marsh grass is now a lacy silhouette as the sun completes its overhead journey for another day. I feel the first stir of an evening breeze . . . maybe rain.

Everyone in the pond has eaten today—the teals, the heron, the alligator. I know that if it were not for the alligators, the ponds would not be kept open—and no one would have eaten.

Still, I'm strangely sad about the lovely little teal.

4.
Crows and Brown Paper Bags

"I didn't know they bathed," the fellow said in disgust as we stood watching a crow doing just that across a shallow pond. A lot of people feel that way about crows. Personally, I like them.

My encounters with crows have convinced me they are more clever than I. For instance, one thing I've discovered is that they know brown paper bags often contain food. There was the egg incident that I remember especially.

One day I bought a dozen eggs in a small store in the town of Flamingo and headed back along the bay to my tent. I decided to sit by the shore for a little while and watch the pelicans diving. A huge crow hunkered in a nearby tree, watching me watch the pelicans.

I soon forgot about my brown paper bag. The crow didn't. Stealthily, he dropped down and ripped open the bag, pecked away the plastic carton, and feasted upon my breakfast.

26

When I discovered the thief at work, it was too late. In spite of the ranting, raving human coming at him, the crow calmly jabbed the last egg with his stout bill and then alternately waddled and flapped off, head held high, egg stuck neatly on the end of his beak. I soon gave up the chase and went eggless that day. I have since gone sardineless, appleless, grapefruitless, raisinless, and doughnutless.

Another time, I was canoeing on the Buttonwood Canal and had just tied up to have lunch on a can of sardines, which, of course, I had in a brown paper bag. Before I could take one bite, I was joined by a very handsome crow. I introduced myself and offered him some lunch—I had long since forgiven the egg thief. He watched me from his perch twenty feet away, *chacked,* and dropped lightly down to a branch six feet from me. He cocked his head and looked piercingly at me with one eye. He hopped closer.

I held out a tempting morsel, hoping to teach him to take it from my hand. He glared and stoutly tapped the branch with his bill. I again offered a hand-held tidbit, assuming his tapping was nervousness or a displacement activity of some sort.

Again he tapped in the same spot with much gusto. His head cocked first one way, then the other, one intense eye looking at me with each turn. He tapped furiously now, losing patience with me. Again I offered. Again he tapped. Would he never understand what I wanted him to do?

Would I never understand what *he* wanted *me* to do? I stretched forward as far as I could and placed the sardine on the exact spot he had demanded. He snatched it and was off without so much as a backward glance.

5.
The Eagle and the Osprey

I remember hearing an osprey one time up on the Shark River. Far above, the osprey and a bald eagle circled. The osprey had a fish. The eagle wanted it. So the aerial dog fight—or bird fight—began.

The two dove as one, first left, then right. The osprey did an Immelmann, made a sharp turn, and dove. The eagle followed like a shadow, immense wings shining golden in the sun, immaculate white tail fanned.

The osprey dropped the fish—why, I don't know. But the fish fell and was reclaimed by the river. Both birds hovered above the spot. Then, incredibly, the osprey attacked the eagle! The eagle tried to take his leave. The osprey was adamant. Shrieking and haranguing, he gave his tormentor no peace. Only fair, I thought. The two disappeared over the treeline but I could still hear the osprey's angry, scolding cries.

Lewis '85   Brown Pelican

6.
The Snake and the Hawk

Red-shouldered hawks are very common in the Everglades. You can't do anything there without seeing several a day, usually in pairs. They are, for the most part, very tolerant of humans.

One day I come to realize just how tolerant they are. I stop my canoe by a large open pond where some white ibis are feeding. In the center of the pond is a dead tree thirty feet high. On its tip, not surprisingly, is a red-shouldered hawk. He looks directly at me. All birds of prey seem to have an angry look about them, and he is no exception.

He has no intention of leaving. In fact, I have the curious feeling that he has a purpose for staying. Suddenly, there is a terrific *whooshing* sound directly behind me. I almost fall out of the canoe in my haste to turn around.

Locked in a death struggle on the half-submerged, finger-like roots of the black mangrove

are the hawk's mate and a large yellow rat snake. The female had been perched on a branch just above and behind me, waiting, ignoring me, intent on the business at hand. Her wings are outstretched now, her hooked bill agape, awesome talons grasping the snake's middle.

The snake thrashes head and tail ends wildly. The hawk pushes her wings forward, forcing herself and her writhing prey backward toward solid ground. A great effort and the struggling pair disappears, flapping and crashing into the dense gloom of the mangroves. A moment later they reappear in a patch of dust-laden sunlight.

The hawk has reached dry ground with her prey. A mighty push and she is aloft, the snake, a mythical flying serpent, still convulsing in her talons.

I turn to look at her mate atop the dead tree. He screams . . . waits. There is an answering scream. With one last look at me, he too is gone.

Marsh Scene, Everglades 1975— Levin

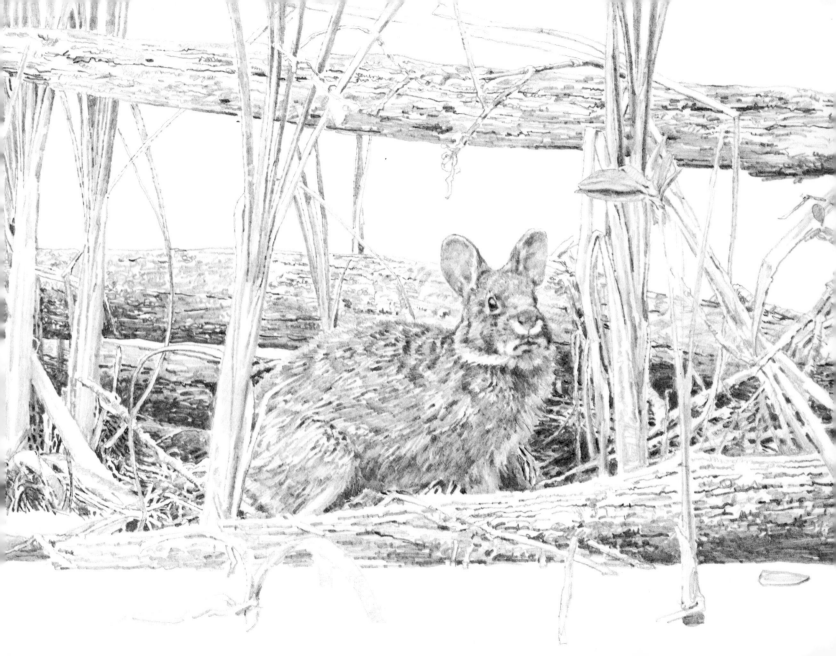

7.
Temple of the Totems

Silence.

The first pale amber light of morning has not yet appeared. In the marshy clearing, all is quiet. Against the gray dawn sky are etched tall, lifeless columns. Atop each one rest mystical, feathered demons, sometimes singly, sometimes in pairs. They wait, still as death.

The amber light appears, wan at first but increasing quickly in intensity. It glints through the leafy walls of the temple, radiating cross hairs of light.

With the light comes warmth. The black demons awake. High on their gnarled columns, they have not yet been touched by the rays. Slowly, a few lift monstrous wings away from their bodies.

The light turns gold. More and more of the giants lift their wings, and the gold slowly

plays on them as it clears the surrounding forest. It reveals gargoyle heads, some flame red and naked, others gray and candled. All have huge, hooked beaks.

As the warmth cascades over their backs, the demons stretch their wings as far as possible. The wings are iridescent, first blue, then gold. The great primary feathers spread like the fingers of a colossus.

Bathed in light and warmth now, the demons are tense and rigid, absorbing as much as they can.

Then, release. The first has broken free from his column, taken wing. Then another, and another. The light is blinding, broken only by the passing of the blue-black bodies.

They are gone. The vacant pedestals remain, merely dead trees in the marsh.

The day has begun.

Turkey Vulture
Everglades
...

8.
Bear Lake Trail

Bear Lake Trail is really an old canal, dug in the 1920's. Abandoned soon after, you would never guess its origin from the way it looks now. Of course, down on the coastal prairie, the long, straight stretches give it away. But at the trail head it looks as if it had always been here. Everything grows in great profusion along the banks—cactus, palm trees, mangroves, bromeliads, and gumbo limbo trees, their sunburned bark hanging in shreds.

I launch my canoe in the early morning. The branches overhead, reflecting into the opaque water below, create a shimmering, brown-green tunnel. All is still, without movement or wind. Only the mosquitoes are busy.

My first event is a barred owl on a low-hanging branch, spotlighted by a ray of sunlight filtering through the tunneled ceiling. The owl's huge brown eyes are unafraid. I slide silently beneath him.

"Aristotle"
Barred Owl hit by
car and adopted
by Park Ranger.
Blind in right eye

A loud squawk is followed by irritated *cluck-clucks* . . . a little green heron, disapproving, and always just ahead. So it will be all day with little green herons.

Drifting . . . suddenly, before me, a fairy dance of huge snowflakes, reflecting, rising, tiptoeing on the water. Egrets. Their graceful dance is not for me, though, but for the fish below. As their tiny quarry flee in panic, they are snatched up in the dancers' sharp bills. As if by a signal—given by me, I fear—the egrets disappear in a flurry of hoarse squawks . . . homely sounds for such lovely ballerinas.

It is still early morning but very hot. The next turn of the canal brings me to an open space, a large shallow pond filled with snags and dead trees. The pond is quite still. Feeding at the far edge, among the mangrove roots, are three white ibis. One youngster still wears his marble-cake colors. Soon he'll be as white as the others and his long, down-curving bill will be as red as lipstick.

There are Louisiana herons, and a great blue, and I see my dancers are here also. At the top of the highest dead tree sits a red-shouldered hawk. He glares down his bill at me. He is very pale for his kind—most beautiful. The scene is idyllic. I decide to leave it so, and quietly slip away.

Occasionally the water wells up and explodes at the passing of the canoe. Catfish? Tarpon, perhaps. A hundred yards ahead I see ripples on the surface of the water and feel a welcome breeze. After the closeness of the tunnel, Bear Lake looks like an inland sea. Its shores are lined with the bleached skeletons of a thousand trees, killed in a great hurricane.

Overhead, a high-pitched whistle, and an osprey hovers momentarily. He dives from

a hundred feet or more. At fifty feet he corrects his angle of descent and drops like a rock. A great splash and he's completely submerged. His head reappears. With tremendous effort, his marvelous wings free themselves of the water and thrust him forward and up with a sea trout. His prize is so heavy he begins losing altitude. More power is needed. He has it, and is free. Gaining steadily, he disappears over the treetops.

Gator Lake and the bay still beckon and I push on. The tunnel swallows me up again. Alongside the canoe, a dark torpedo shape keeps pace—a garfish. He leaves me for the privacy of the mangrove roots. An incessant rattling breaks the solitude, and a kingfisher disappears down the tunnel. A little green heron leaps across the trail and skulks off, grumbling under his breath.

Soon, Gator Lake—not much of a lake compared to Bear Lake, just an open marshy area. A great rustling sound and rush of wings. I've surprised several turkey vultures dozing in the leafy recesses of the mangroves.

Overhead, a wondrous flight of white pelicans. Maybe twenty-five or thirty of them on a thermal, spiraling effortlessly upward, higher and higher until they disappear in the white of a thunderhead.

The trail turns now and becomes quite wide and much more open. There is a bit of wind as I enter the coastal prairie—mangrove islands everywhere, and brackish marshes. I realize I'm beginning to tire. I've paddled nine or ten miles.

A shadow darkens the sky. A huge black-and-white form glides over. Then another, and another. Wood storks—the largest of the wading birds, almost as tall as a man with a wingspan that seems endless. Fatigue is forgotten. Walls of green enclose me. Unlike

53

the tunnel, these have no ceilings. I drift forward . . . a lone pelican sits on his reflection in the brown water, looking for all the world like a Viking ship. He leaves the water in a bound and settles on a slender branch that supports him easily. He must be very light in spite of his great size.

More green islands—distant shores—then, incredibly . . . pink. *Pink* pink. The pinkest pink. Spoonbills wading in the shallows . . . their heads sweeping back and forth, great, flat spoons probing the mud below for its life. An open area of water separates us, too shallow for the canoe. It is just as well. They are very shy and would only fly away. While the canoe settles in the muck, I fill my eyes and soul with pink.

The wind picks up, reminding me that the bay is waiting. My map tells me there is a sharp turn ahead. What the map doesn't tell me is that the shoreline at this point will be a brawling, preening, feathered wonder. Everywhere, white pelicans, white ibis, egrets, wood storks, and, off a little to themselves, my spoonbills.

A right angle bend, and the trail is as straight as a bowling alley. At last I reach the dam, a giant plug, naked and faintly ominous. I beach the canoe and step out onto the marl. A mixture of clay and calcium carbonate, it is bone hard when dry, and ankle deep muck when wet. Hundreds of portages have created a small tidal inlet. Huge blue crabs snap at me in anger and slip back into the ooze. Beyond the dam is a wide tidal river and azure-blue bay and azure-blue sky, all one.

White Ibis
Bear Lake - Everglades -
penni 1975

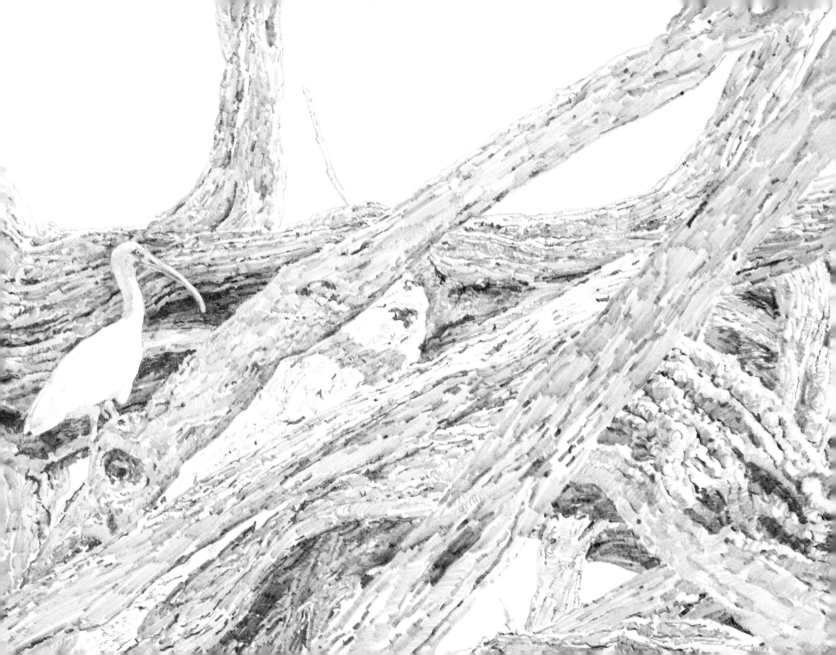

9.
Journey's End

Florida Bay is where all journeys in the Everglades end and where so much life begins.

I glide in my canoe on placid blue-green water. To the north is an endless wall of mangrove trees; the southern horizon is punctuated by tree-islands. Stilt like, these trees stand as fragile buffers against the nightmare force of the hurricanes that howl out of the sea, hurling giant sea turtles twenty miles inland and flattening birds like splashes of paint on tree trunks.

The bay is quiet now—but beneath its surface is a stew of life. A great silver tarpon lifts himself from the water and hangs like a trophy in the air before slamming back into the sea. Above, ospreys patrol, hover, dive, and come away rewarded. Their screams, echoing along the tree walls and across the shimmering water, carry the sound of heartbreak. There are crocodiles, too. Not many now, and perhaps none soon.

Mangrove
Everglades 1975
Lewis

The wind has come up suddenly from the southeast. The bay is ridged and the color of iron. I put my back to the wind like a sail and plane on the following seas. Soon, a break in the green façade. Ahead is East Cape, a white sand crescent curving off as far as the eye can see. Around the bend the wind dies and the water is peaceful. I beach the canoe and stand beneath a single coconut palm. Just above the high-tide line, spiky yuccas form a living chevaux-de-frise, protecting the forest beyond. The horizon has blended with the sky in the shimmer, giving me a strange feeling of vertigo.

Half buried in the sand, like long forgotten figureheads, are the great trees killed by past hurricanes. The beach is silent and hot. The only sound is made by my feet crunching the coarse shell sand. I see a fin break the surface glare off shore. A porpoise makes a slow, deliberate arch, then disappears, only to reappear in a moment.

It is late afternoon. In the red, slanting rays of the sun, a scissor-tailed flycatcher perches in a tree. Using his long tail feathers deftly, he launches himself from the topmost branch. High in the air, he snatches an insect on the wing, dives, and makes a spectacular pull-out at his perch.

Thunderheads have filled the sky, and the setting sun sends silver rays through them, creating awesome, celestial cathedrals. Mere inches above the water, flights of brown pelican silhouettes glide by, riding a thin air cushion. The pelicans are heading to Flamingo for a handout from fishermen cleaning the day's catch. The sun balances neatly on the rim of the bay, and in an instant melts, and is gone.

$\mathscr{Index}$

Page numbers in **boldface** refer to illustrations.

64